CONVERSATIONS WITH CAT

An Uncommon Catalog of Feline Wisdom

CONVERSATIONS WITH CAT

An Uncommon Catalog of Feline Wisdom

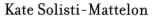

Kate Solisti-Mattelon

BEYOND
WORDS
Publishing
I N C

Beyond Words Publishing, Inc.
20827 N.W. Cornell Road, Suite 500
Hillsboro, Oregon 97124-9808
503-531-8700
1-800-284-9673

Editor: Laura Carlsmith
Managing editor: Julie Steigerwaldt
Proofreader: Marvin Moore
Interior design: Principia Graphica
Cover design: Dorral Lukas
Composition: William H. Brunson Typography Services

Printed in Malaysia
Distributed to the book trade by Publishers Group West

Library of Congress Cataloging-in-Publication Data
Solisti-Mattelon, Kate.
     Conversations with cat : an uncommon catalog of feline wisdom /
Kate Solisti-Mattelon.
        p.   cm.
     ISBN 1-58270-062-1 (pbk.)
     1. Cats—Behavior—Miscellanea.   2. Human-animal
communication—Miscellanea.   I. Title.
SF446.5 .S66   2001
636.8'01'9—dc21

                                                          2001037464

The corporate mission of Beyond Words Publishing, Inc.:
Inspire to Integrity

CONTENTS

Acknowledgments VII

❧

Introduction IX

❧

Chapter One:
Where Do We Look for Answers to Our Questions? 1

❧

Chapter Two:
Earthly Beings with Cosmic Understanding 5

❧

Chapter Three:
Life According to Cat 17

❧

In Conclusion 131

❧

About the Author 133

Acknowledgments

I am deeply grateful to the Council of Felines for sharing their wisdom through me. It's an honor to be a voice for such extraordinary beings.

Thanks to all the people who sent questions for the Cat Council to answer: Alex Mattelon and Miranda Solisti, Dana Barker, Bev Bucklew and other staff members at Best Friends Animal Sanctuary, Marni Charniss, Gail Corte, Lila Devi, Leslie Engleman, Jill Fineberg, Joni Gang, Dorie McCubbrey, Linda Neu, and Pam Wood.

Special thank yous to Sue Green of The Whole Cat for her wonderful photo and ongoing support and to Anita Aurit, president of ThePetProject.com, for her fabulous review.

Thanks to Cindy Black and Richard Cohn and their fabulous, cheerful staff at Beyond Words, to Laura Carlsmith for her sensitive editing, to Julie Steigerwaldt for typing in all the changes, and to Marvin Moore for his careful proofing. Thank you all!

Thank you Patrice for your love and support and for sharing this amazing life with me. Every year is better than the last!

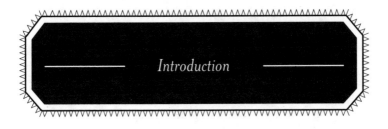

Introduction

Since humans first began living in small groups, animals have been our companions. Sometime, thousands of years ago in a village on the edge of a desert, a small feline chose to begin hunting poisonous snakes that threatened the village inhabitants. People marveled at this being's ability to move easily among humans yet to maintain her wildness. Why did she leave her world for the human village? Why did she stay?

Cats have been challenging the human mind and heart ever since. Nobody seems to be neutral about cats. We're either positively poetic or menacingly negative. Throughout our shared history, cats have inspired artists, writers, poets, and composers. Cats were worshipped in ancient Egypt for thousands of years only to be visciously persecuted in medieval Europe. What is it about this little animal that brings out such extreme emotions in human beings?

In this book, we hear from the cats themselves. With their sometimes surprising answers to our questions, perhaps at last we will begin to understand some of the mystery and mystique that surrounds the cat.

Felis catus, our domestic feline, is descended from the little desert cat who walked into the first settlements along Egypt's Nile River. How did this ferocious hunter come to live with us? Who is this animal? Are they really our companions?

Through telepathic communication, cats told me once, "We weren't worshipped in Egypt just because we were pretty and good mousers!" There was something more. Something energetic and spiritual, something extraordinary that cats did for and with people.

What do your cats bring to you?

Why *Conversations with Cat*? Today, people are hungry to understand the whys and hows of our lives. There are countless self-help books about learning from esteemed teachers, on connecting to our own souls, and on listening to God. Wouldn't it be amazing to accept that the furry beings sleeping on our beds, curled up on our laps, and weaving in between our feet are capable of teaching us everything we are seeking? We simply need to open ourselves to hearing what they have to say.

Communication with cats, as described in this book, goes beyond the parameters of feline brain size, physical

capabilities, instincts, and behavior. I am not interested in proving that feline consciousness as I've experienced it is measurable by human standards and technologies. This book is about tapping into the Divine Consciousness that operates in and through every creature, plant, stone, and body of water on our planet.

Each unique species and individual expresses Divine Consciousness in his or her own way. I have not interviewed individual cats for this book. Rather, I've connected to the Council of Felines, or higher group consciousness of all cats, in order to get the "big picture" as opposed to individual opinions. In this book, I approach cats through this shared spiritual connection. In this place of unity, there is no need for spoken language. Divine Consciousness is expressed as a language of the heart, as interspecies communication and understanding. The information in this book is not definitive for all felines. I am a vehicle for this information, and I have my own filter system. I'm only as good a receiver as my consciousness level at this moment allows. As with most information we humans share with one another, the final proof of its usefulness comes through testing it against our own life experiences.

Conversations with Cat introduces readers to the spiritual, physical, emotional, and mental awareness inherent in the feline species. Individual levels of awareness vary from cat to cat, just as they do from person to person. This book is not a cat-care manual. It is a tool for deeper understanding.

Chapter 1 invites readers to open their hearts and minds to communicating with cats.

Chapter 2 examines why we can look to cats to help us understand life.

Chapter 3 offers feline answers to human questions.

It is my hope that books such as *Conversations with Cat* and *Conversations with Dog* can help people remember how to tap into the Divine Consciousness in all of creation—animal, vegetable, and mineral—and so share our experiences once again as ancient legends say it was intended. I will continue to use my gift of interspecies communication in order to deepen understanding and connection between all beings. For me, it's all about love.

Our ancestors respected animals as healers and teachers. When the human race was younger, we learned and received many gifts from species other than our own. We recognized that our interdependency created balance and life. But as time passed, people began losing this connection with other species and with Mother Earth. We headed down a path of self-centered, myopic existence focused on accumulation and domination. As a result, today, much of

humanity feels discontented, purposeless, unhappy, unfulfilled, fragmented, and alone.

There are stories in every human culture about the deep connection we once enjoyed with the natural world. There are also stories of how we became separate and isolated from other species and from Mother Earth as well. I received such a story, but in it three species choose to remain with us to help us remember who we truly are. Here is the story:

Once, in the beginning of time, there existed a council of beings, each of whom represented different expressions of the Creator in all sorts of marvelous forms. At the council table sat representatives from the insect, reptile, bird, mammal, marsupial, and human families. Each shared his story about how his species incarnated to experience life in a particular form in order to learn and to share specific truths with all other beings. The common goal was to better understand—together—ourselves and our Creator. Ant chose a form that would help him learn to cooperate with the plants. Humpback whale chose to be a singer and to explore the interaction of sound and water. Falcon chose to experience rapid flight and to explore the element of air. Humans chose to explore our remarkable intellect, manual dexterity, and ability for spoken language.

In the beginning, every being was connected heart to heart as one family. They enjoyed sharing with each other their new experiences of life in their chosen form. But one day, in the pursuit of the mind, and as our attachment to our own accomplishments grew all-important, we humans stopped returning to the council fire. We became self-absorbed, forgetting that we had agreed to share our experiences with other beings. We began to feel superior; we began to forget that we were all one family. A chasm between humans and other creatures was created. Most of the other animals continued on their own paths, but three species lingered, determined to mend the ever-widening gap between humans and other creatures. At a critical moment, they consciously chose to leave the comfort of their fellow creatures, to a great extent, and to accompany the humans, hoping to lead us back to the Creator, back to connection, back to Love. These three species were the cat, the dog, and the horse. To this day, we have only to stop and pay attention and these three will remind us of who we truly are.

So, now, in this special moment in time, we can stop and ask cats who we are and what they are up to with us. What an opportunity!

When I wrote *Conversations with Dog*, I was deeply touched by canine wisdom, with its constant gentle compassion for us human beings. As I sat down to talk with cats, an entirely different energy emerged. Cats tell it like it is! Cat people will thoroughly chuckle at the cats' sometimes wry and succinct answers to our questions. Read together, *Conversations with Dog* and *Conversations with Cat* shed light on these extraordinary beings who love us in their very different ways! How truly blessed we are to live in the company of animals. Enjoy!

"WHAT SORT OF PHILOSOPHERS ARE WE, WHO KNOW
NOTHING OF THE ORIGIN AND DESTINY OF CATS?"

—Henry David Thoreau

T hroughout history, humans have been graced with
extraordinary human teachers, masters, saints, and sages.
But no matter how brilliant or evolved these human
teachers may have been, they were still human. How many
of them understood the experiences and perspectives of
other species?

Today, we humans understand the importance of
learning other people's languages and customs if we are to
communicate beyond superficial levels. Yet most of us
don't even consider the possibility of learning another
species' language or culture! But what if we could learn
the language of animals as easily as we learn another

human language? Wouldn't that broaden our perspectives and understanding of the world in dramatic new ways? What could we learn if we opened our hearts and minds to other non-human realities? Perhaps it's possible for us to find our sacred humanity through understanding animals. Perhaps, by understanding how animals perceive us and the world we live in, we'll learn more about ourselves and our relationship to all animals and to our planet.

Think about it. If you love animals or live with them, you experience interspecies communication all the time. Look at your cat: an arched back, a happy purr, wild eyes, a meow at the door, rubbing against your leg, a kiss on the nose. You understand what she is saying. You may even understand more subtle, personal signals, such as a particular movement of a tail or a paw posed to box your cheek to let you know you're out of line. Start paying more attention to the unique signals that your cat sends you, and you will begin to see how good animals really are at letting us know their needs.

Do you speak to your cat in a soft, sweet voice, creating a special language between the two of you? What do you see in your cat that is different from other cats you've known or lived with?

Now that you're thinking about how much you and your cat really do understand each other, let's go to the next level. Observe your cat with this new understanding that you two are communicating all the time.

The first thing most people do with friendly cats is touch them. Their beautiful, soft fur gives us instant pleasure, and the cat takes pleasure in our touch, sensuously rising to meet our hand, turning different parts of his face and body into our hand. Then a sound rises deep from within the cat—a sound so loving and gentle, we want to hear it as often as possible. We feel so special when a cat jumps on our lap, settling down to be stroked, blessing us with his purr. Does the cat know the effect he's having on us? Does he know how he makes us feel? What do you think?

From now on, when you touch your cat, be aware of how you are touching him. How does he respond to different strokes, scratches, pressures, speeds? What happens when you leave your hand just above his body? Can you see how he feels your hand even though you aren't making physical contact? By closely observing your cat and his responses to you, you are learning to pay attention to subtle communication. With practice, this observation will develop into deeper awareness. Conscious touch with clear, simple communication can lead you to develop your connection with your cat into a beautiful partnership. Love combined with all of the above will lead you into understanding yourself and the essence of Cat.

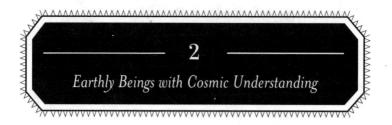

"CATS ARE WORKS OF ART. IN A CAT, YOU HAVE A
WONDERFUL WORK OF ART THAT WILL ACTUALLY LIVE WITH
YOU, SIT IN YOUR LAP, AND PURR."

—Jane Bryant Quinn

Cats have never lost or forgotten their connection to God.
For centuries, spiritual teachers, masters, and beloved
humans such as Jesus, Buddha, and Mohammed have told
us that we are one with God. We still struggle to believe it.
Animals never struggle with this. They *know*.

Like us, cats experience physical reality: pleasure, pain,
hormonal impulses, joy, and sorrow. Unlike us, however,
cats know and live every moment of every day in a state of
oneness. Do cats *know* that they know? Of course they do!
Let me share my story of what my own cat taught me.

His name was Dusty. I hope his story will touch those
of you who have had a similar experience and help you

remember when you had a deep connection to an animal or to the natural world. Perhaps remembering will help you heal painful memories and allow you to reconnect and open your heart again.

I was three years old and had been "hearing" animals since the beginning. I "heard" their thoughts and feelings. I could send my thoughts and feelings back to them. Not only could I have conversations with animals, but I could share thoughts with plants and trees. This seemed normal to me, but I quickly learned that the human beings in my family were not having the same experiences. As I read their thoughts, I found that they had a sort of big door separating their hearts and minds from the plants and animals—and each other. I didn't understand why they had this big door and I didn't. But, being very young, I couldn't ask. I just started sharing my experiences, expecting that I would receive love, understanding, and acceptance.

At first my parents smiled and told me that my conversations with roses and insects were just my imagination. Then one day, during a tea party my mother was having with some friends, I listened to what the women were thinking, not to what they were saying. I was struck with powerful emotions. After the party, I told my mother that Mrs. H. really hated her husband, since this is what I had "heard" from her mind/body. My mother turned on me, saying, "That's not true! Where did you get such a crazy idea?" She wasn't waiting for an answer. She was telling me I was bad and wrong, or so I felt. My mother and father began to look at me funny. They began to worry about me. I was confused. I began to worry that I was different. I stopped sharing the conversations I was having with the plants and animals.

Then in November, when I was three years old, my father surprised me with a little orange tabby kitten. It was love at first sight. Dusty began speaking to me in the language of thought right away. He told me that he'd come to be with me, to share my experiences with the plants, trees, birds, reptiles, and insects and that it would be better if I didn't tell my parents anymore. He said that they weren't open in the same way that he and I were. I was relieved and a little sad but delighted that Dusty would be with me.

Dusty and I were inseparable. He attended tea parties with his own saucer of milk. He let me dress him in doll clothes occasionally and drive him around in a baby carriage. He slept in my bed and one evening jumped in the bath with me. I'll never forget the look on his face when he hit the water. He was mortified and did his best to get out and walk away with dignity. Adults noticed how close we were, commenting that Dusty was more like a dog than a cat.

Our most extraordinary times were in the garden. We'd sit and talk with all manner of plants and creatures. I learned how roses have a love affair with the sun and how turtles can predict the weather. I asked Dusty questions, and he always answered. These were not just conversations between a girl and a cat. These were questions and answers between two souls, souls currently inhabiting a girl's body and a cat's. We often felt as if we were one being. This same feeling of oneness came over me whenever I connected to the other animals and plants. I was happy, really happy, surrounded by unconditional love from animals and plants and Nature herself.

When I turned five, it was time to go to kindergarten. I don't remember too much about the other children except that I had to communicate with them through spoken words only. They, like the adults in my world, couldn't hear my thoughts. They'd already learned to close their minds in order to fit in. Or perhaps they'd never been as open as I was. The greatest joy in my life continued to be spending time with Dusty.

When I started first grade, Dusty told me that it was time for me to really delve into school. He told me that I

Sure enough, when I got home from school, my parents told me that they had found Dusty by the side of the road, hit by a car. My father handed me Dusty's beautiful blue collar with one of the bells squashed. "Where is he?" I cried. "I want to see him." My father replied, "I buried him behind the playhouse. Come, let's go see him." Still numb, I walked hand in hand with my father to Dusty's grave. My father had fashioned a wooden cross for him, and I hammered it into the ground. At last I was overcome. Tears streamed down my face. How could this happen? How could he die? My father tried to console me, but nothing could stop my sobs. For days I grieved. My friends couldn't help me. My parents couldn't reach me. I felt utterly alone. I tried to talk to Dusty, but I could no longer hear him.

Soon after, I got tonsillitis; over the course of a few months, I got sicker and sicker and I wanted to die. My mother commented on how the illness had changed me from a sunny child into a dark and sullen one. This change was not due to my illness; it was due to the loss of my best friend. When it was clear that my tonsils had to come out, I was terrified, sure that I would die all by myself and far from home. The hospital was cold and scary. However, as I went under the anesthesia, I felt safe for the first time since Dusty's death. Some presence was loving and protecting me. I felt someone gently pick me up and cradle me, assuring me that I'd be all right. I became aware of lying against warm fur. I smelled a familiar smell. It was Dusty. He told me, "It is not your time. You have work to do. I will always be here." Dusty came to reassure me and to help me make the transition to the next phase of my life.

I woke up to find my mother standing over me. I was OK.

I went home to a doting family. When I was well enough to go outside, everything was different. I deliberately began to shut out my animal and plant friends. I began to build the big door. I did this partly from guilt; I felt that Dusty had died because I turned away from him. I also did it to fit in, to be less sensitive. By the time I was eight, my ability to hear plants and animals was gone. I accepted that I needed to live like the other humans around me—in separation. Now I could be like everybody else. And more importantly, I would never have to feel as deeply as I had before.

When I was twelve, I developed severe allergies and asthma. Tests indicated that I was terribly allergic to cats. On some level my closedown was now complete. I could never again be close to cats. I would be safe now from deep loss and pain.

As I grew up, I never allowed myself to be as close to another cat as I had been to Dusty. In fact, I never allowed any human being to get as close to me, either. As I grew older, I channeled my love of animals and the earth into the environmental movement of the '70s and '80s. When I was living near Washington D.C., working for a national environmental organization, a breakthrough occurred.

I had decided to go to therapy to deal with the usual adolescent anger issues. I was ready to move past the past. After some months of therapy, I discovered that I had separated out a part of myself. I had rejected my intuitive, feminine side, pushing it away in order to be successful in the world. However, this missing part was critical to my healing and becoming a whole person. It held the key to my true, authentic self.

As my feminine, intuitive side began to feel safe enough to join the rest of me, unusual things began to happen to me. On afternoon walks by a nearby stream, I started hearing voices. It seemed that the trees were whispering to me. I was twenty-eight years old, married, in a responsible job that I enjoyed. I knew that I wasn't crazy. But beyond knowing that I was mentally stable, the feelings that washed over me when the trees spoke to me felt beautifully correct. I knew that something this loving could not be bad.

My rational mind had no reference for talking trees. And at that point, I did not remember that I had heard trees when I was young. I did begin to think about the con-versations I'd had with Dusty. Was it possible that I not only spoke with him but with other non-human beings as well?

My only known reference for communicating with animals and plants was the Native Americans. One of the basic tenets of Native American life is the belief that animals, trees, plants, rocks, streams, rivers, and oceans are conscious beings capable of communicating with any-one who takes the time to listen. I started reading. As I read beautiful words about the entire natural world being family to humankind, my heart began to open. I cried. I laughed. I got it! I understood, deep within my being, that it was true: animals and plants can communicate with human beings. I felt wonderful.

Then I got really depressed. I didn't have an ounce of Native American blood in me. I had finally found people who understood what I understood! But I didn't "belong" to these people. How could it be that I felt totally different from my own roots and wanted desperately to be a part of theirs? In a Native culture, my communications with trees would be not only accepted but valued. In my culture, people thought I was nuts! Why could *I* communicate in this way? What was I supposed to *do* with this gift? I decided to keep it mostly to myself, since my few tries at sharing it had been met with concern and skepticism, but I had to continue to nurture it and learn to understand it better.

I remembered that all of human cultures started with a nature-based religion. Because of my northern European ancestral roots, I decided to join a study group to learn about Celtic and Norse religion and mythology, specifically Treehenge. Like the better-known Druids of Stonehenge, the ancient Norse believed that trees were teachers and could be prayed to for guidance and inspiration. I learned to meditate with trees. I began to experience how focus and concentration helped my gift develop. I also learned that strong emotions clouded what I was able to receive.

I began to feel again what it was like to be connected to every living thing. I understood that we truly are all one. Dusty had provided me a firm foundation in loving and trusting this knowing. If not for him, I would have completely buried my abilities to communicate and might never have regained them or my memories. Dusty was my link to love, to God.

Cat's Teaching Starts with Our Paying Attention

How do we learn to tune in to this love? The first step is to wake up and be present. How do we learn to be present in our

lives every day? We can look to cats. They are fully present in the moment. Notice how a kitten focuses on a string. She isn't thinking, "Gee, it would be nice if I could catch that string." She's thinking, "I've got that string *right now!*"

Most of us live in the past, prisoners of our experiences with parents, schoolteachers, or bosses. Others of us live only in the future: "When will I meet my true love? I'll speak to my boss tomorrow. I'll take the kids on a really great vacation next year. What if I don't save for retirement?"

But when we live in the past or in the future, we become completely disconnected from the present. Life passes us by. We search for happiness instead of creating it now. As long as something is being searched for, worked toward, hoped for, it is never here.

Observe how you speak. Are you repeating a script from the past, or do you mean what you say? Do you use phrases like "I will start saving with my next paycheck" or "I will earn more money someday"? Someday is never today. Take a tip from cat; try shifting the way you speak, incorporating present tense for future tense. Try saying, "I have all the money I need and want right now. I am creating the life of my dreams." Practice enough and the universe will catch up.

To pay attention and to be fully present in life is to use all the gifts you have been given. Use all your senses, including your feelings and intuition. Take time to savor tastes and smells. Stop and really listen. Look at the beauty around you. You can choose the life you'll experience. The cat in your life can be a most valuable teacher. Let your cat show you how to *live*!

"CATS ARE INTENDED TO TEACH US THAT NOT EVERYTHING IN NATURE HAS A FUNCTION."

—Garrison Keillor

The first obstacle to interspecies communication is the belief system that has been programmed into us: that talking with animals is impossible. But remember, other cultures have not accepted this belief system. Indigenous cultures not only believe that humans can communicate with animals, plants, and all of Nature, but they honor and respect this communication as vital to living a balanced life.

You will find that *believing* in interspecies communication takes effort and attention since the collective belief system exerts powerful energy in the opposite direction. Just remember my story. You'll receive the same rewards

I've received from forging ahead and breaking out of the collective trance!

As you break free, make time to sit quietly at a comfortable distance from your resting cat. Some cats will want you right next to them. Others need some space in order to feel that they can continue napping or resting uninterrupted. Observe this space. Notice at what distance your cat accepts your presence and when he feels uncomfortable. Observe your own energy. Are you calm and peaceful or are you eager or excited? A calm and peaceful attitude will be easier for your cat to accept.

Now just sit and observe your cat and yourself. If you're like most of us, sitting still and doing nothing will be challenging. Your mind will start working. You'll start thinking about other things. Gently bring your mind back to observing your cat, the sunshine in the room, the quiet. Each time you feel distracted by your mind or an outside noise, bring your awareness back to your cat and your own breathing. Notice how your cat's sides rise and fall rhythmically. Try to breathe deeply into your own belly in the same rhythmic way. Sitting quietly and being fully present is a way to learn how to *be* like a cat. With practice you will become more and more sensitive to yourself and to your cat. *Being* like your cat, even for just a few minutes

a day, will facilitate and deepen the communication between the two of you.

Getting Started

Now that you're practicing being with and like your cat, what questions would you like to ask her? The best place to start is with the questions I have heard many people ask about cats. The answers on the following pages, on topics both spiritual and mundane, may hold some surprises.

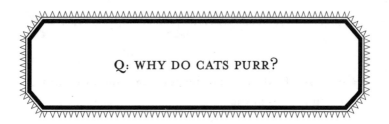

You laugh, we purr.

A: Human beings possess the gift of laughter. Cats possess the gift of purr. Our purr is comforting to us. We purr when we're happy. We purr when we hurt, in order to comfort ourselves. Purring is made up of specific vibrations. These vibrations can heal and recalibrate our bodies and internal systems as well as calm and help heal and recalibrate others, especially humans. You need all the help we can give.

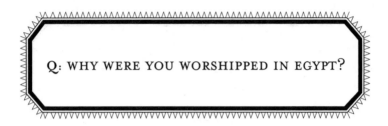

We had a special healing relationship with human beings.

A: We weren't worshipped in Egypt just because we were pretty and good mousers! Our high status came because we had the ability to augment and support the healers and physicians. When a person was sick, we purified the healing room by clearing imbalanced/diseased energies and thought forms of the patient and/or the healer. Then we helped connect the healer to the source of all healing— God. The next thing we did was tune in to the patient and begin clearing him or her of illness and disease in partnership with the doctor.

Q: DO CATS HELP "HEAL" US TODAY?

⌘

Absolutely.

A: The cat in your home has the same ability to help you heal as her Egyptian ancestors had. For example, if you have the flu, we can read the organs and systems to pinpoint the virus. The virus carries a different energetic vibration than your body's health state. We zero in on the flu vibration and isolate it. Next, if you're ready to release the virus, we suck it out of your body into your aura or energy field. We then take it from your field into ours, where we break it up and release it in an "unharmful" state.

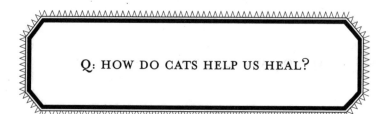

Q: HOW DO CATS HELP US HEAL?

By transmuting your negative energy.

A: When you are ill, there is an imbalance in your energy field or aura. The illness in both the body and the energy field must be addressed together or total healing will not take place. Every night when you go to bed, we work to clear your energy field of imbalances and unnecessary carry-ons.

Every day, many of you accumulate stuff in your energy fields—remnants of emotion, destructive thoughts, and others people's stuff too. The more sensitive and empathic you are, the more stuff you tend to pick up and carry. Most humans are not aware that they do this, although many of you feel "heavy," "tired," or "burdened" after you've spent time in traffic, at the mall, in meetings, or around large groups of people. Your feelings are valid. Taking on other people's energy or stuff will make you tired and sick if you don't get rid of it. This is where we come in. Just as in ancient Egypt, we sense the stuff in your energy field, take it into our own energy field, and then release it. You get cleaned up, feel lighter, sleep better, and can face the morning with renewed energy.

*Your aura or energy field is as obvious to us as
your arms and legs!*

A: When we look at you, we see your physical body, but we also see your aura or energy field. In fact, the state of your aura is often more interesting to us than what's happening with your physical body. When we focus on your aura, we receive a full picture of your physical, emotional, and spiritual health. We literally read you like an open book. This enables us to rapidly judge whether you will be kind to us, dislike us, be afraid of us, or try to approach us. We can then make a quick decision about how to respond to you.

Q: WHERE DID YOU DEVELOP THIS ABILITY?

❧

It's a part of our equipment.

A: We are perhaps the most finely tuned predators on the planet. The cat-family members, in all our sizes, are famous hunters and survivors or we wouldn't inhabit every climate zone on earth. We are extremely gifted at seeing and sensing. Our six senses—sight, sound, touch, taste, smell, and extrasensory perception—are highly refined. We are able to pick up the subtle fluctuations in temperature and/or vibration that indicate a prey animal is nearby.

We use our own energy field or aura to accomplish many things. For example, we hold our aura very close to our physical bodies while we stalk prey. Then, when we come within striking distance, we throw it over our prey like a net to disorient the prey long enough for us to catch it. We can extend our energy field out as a territorial boundary, which also serves as an early warning antenna if prey or fellow predators come in contact with it.

Over centuries we have refined this process in order to survive. When we decided to live with humans, we adapted our personal energy work to aid and support humanity in addition to using it to help ourselves.

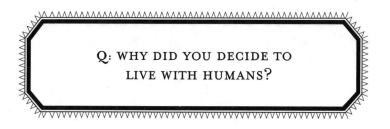

Q: WHY DID YOU DECIDE TO LIVE WITH HUMANS?

For our mutual benefit and spiritual growth.

A: Back thousands of years ago, we chose to inhabit the early Egyptian settlements because, in them, the snakes and mice were plentiful. Also, we chose to help human beings remember who you really are. You are not what you think you are. You are beings of infinite light and beautiful energy. As you have moved out of friendship with the natural world, you've forgotten that. We joined you on your journey to remind you that you are a part of nature, not separate from it. We joined you on your journey to help heal your spirits and bodies and to support you in being whole beings—divine energy in physical form. We joined you on your journey to help you reconnect to God.

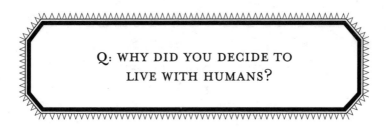

Q: WHY DID YOU DECIDE TO LIVE WITH HUMANS?

For our mutual benefit and spiritual growth.

A: Back thousands of years ago, we chose to inhabit the early Egyptian settlements because, in them, the snakes and mice were plentiful. Also, we chose to help human beings remember who you really are. You are not what you think you are. You are beings of infinite light and beautiful energy. As you have moved out of friendship with the natural world, you've forgotten that. We joined you on your journey to remind you that you are a part of nature, not separate from it. We joined you on your journey to help heal your spirits and bodies and to support you in being whole beings—divine energy in physical form. We joined you on your journey to help you reconnect to God.

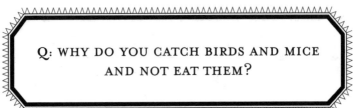

We're not hungry.

A: Nature designed us to be highly stimulated by small creatures that run or fly. In the wild, catching prey is critical to our survival, and the only purpose is for food. Even if we live in a home with a human, we enjoy hunting. It's thrilling. Prey animals understand that their purpose in life is to have babies, nurture the plants, and feed predators. We don't love killing, but we do love hunting, because it's part of who we are. But killing is necessary only if one needs to eat. A cat will hunt for the joy of it, but if we're not hungry, we will not eat, and the prey animal "goes to waste." It is better for both predator and prey if the prey is eaten. If we are killing but not eating our prey, we are out of balance with our environment. If you wish to help us, ask us to eat the prey we kill or stop us from killing.

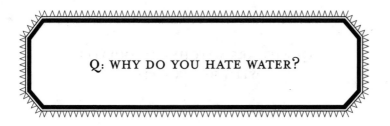

Water is foreign to us at a cellular level.

A: For our desert-dwelling cat ancestor, large quantities of water were unavailable and unknown. We hold a cellular memory of enjoying dripping water, because that was mostly all that was available to us. To this day we prefer to lick dripping water than to drink still water in a bowl. In fact, we are designed to receive our needed moisture from our prey. When we're healthy and eating meat, we rarely need to drink water. And let's not even speak about baths!

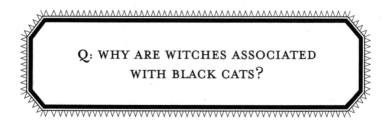

Q: WHY ARE WITCHES ASSOCIATED WITH BLACK CATS?

❧

Bad press.

A: Beginning in ancient Egypt, we have continued helping healers heal their patients for centuries. In ancient times, there were equal numbers of male and female healers in order to offer a balance. As time went on, more women than men stayed on in the healing arts. Men became more interested in the material world and its seductions. Women continued to care for new mothers, children, and the elderly, sick, and infirm. We helped these "wise women" work with the plant kingdom, introducing them to powerful plant healers and helpers such as chamomile and the mint family, herbs for easing childbirth, for pain and purging. These women lived in areas where they could pick or grow plants. They developed their inner senses and were not too involved in secular activities. Their profession and avocation were to connect human beings to the healing available from the natural world. Very often these women lived with a cat or cats as partners in the healing.

As villages grew into towns and cities, these wise women tended to live in the woods, where they were the least distracted. Because their art took time and effort to understand, some people began to fear them. Misunderstanding always leads to fear. People's priorities began to distort as material wealth and prestige took precedence over health and well-being. Cooperation was replaced with competition. Powerful men began to manipulate people so they could control them. Wars raged. Fear ruled. Paranoia followed fear. Everybody was a potential enemy instead of a friend. Life became so confused by the Middle Ages that people actually turned against the very people who remained sane and connected to reality. They turned against the women who brought them into the world, cared for them, healed them, and gave them life. They said that these wise women must be evil and needed to be eliminated. They called them witches and said that they practiced "black arts." Because night was black and everybody feared for his or her life at night, anything black was seen as evil.

But for us cats, night was our friend, the ideal time for hunting. We have eyes that allow us to see rather well at night. We were labeled as "black creatures of the night," evil in every way. Thousands of us, living in what is now Europe, were caught, burned, or hanged along with our friends and partners, the wise women or witches.

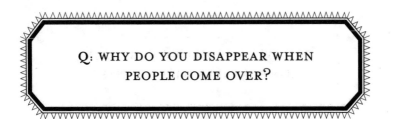

Q: WHY DO YOU DISAPPEAR WHEN PEOPLE COME OVER?

Challenging energy.

A: Because we are gifted hunters, we are extremely sensitive to the way energy flows in our environment. We are constantly adjusting to the energy changes in the home we share with you. We feel your fluctuating energies and have to balance ourselves to compensate for your moods and states of being. When new people come over, we are bombarded with their moods and energy patterns. Some of us have no trouble adjusting to these different energies. Others of us struggle to find our balance, and in order to minimize the impact to our delicate systems, we often choose to avoid new people, especially if they will be in our environment only for a short time.

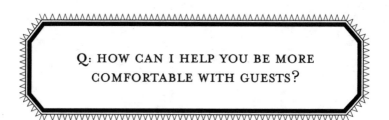

Ask the guests to be quiet and respectful.

A: Most people enter a room with a blast of energy. This helps them get noticed by other humans and make a big impression. For a cat, this is an offensive way to enter another's space. This attitude pushes one's energy on everyone else, whether they like it or not! If a person comes into a room gently and with an attitude of respect for the space, the existing flow of energy is not distorted or strongly impacted. We can feel safe and respected as well.

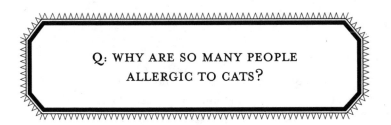

Q: WHY ARE SO MANY PEOPLE ALLERGIC TO CATS?

On some level they may be rejecting life.

A: We are healers, as we've mentioned before. We are fully connected to God. We carry within our cells the memory of perfection. Many of you are afraid that you are not connected to God. Many of you are afraid to live life fully because you were treated badly when you were children. Asthma is an example of how a body reacts to emotional or physical suppression. If a child is emotionally squashed, asthma can be one reaction. In other words, the body internalizes outside pressure into inside pressure. No space to breathe on the outside equals no space to breathe on the inside. When you free yourself of the suppression or the memory of it, you allow yourself to breathe—and to be more alive. We feel pain as you do, but we rarely choose to let pain and misery rule our lives. We strive to live life as fully as possible. That's the energetic answer.

On a physical level, toxins in our environment, in our food, and in our bodies must be processed and released if possible. Many of us release toxins through our skin. Because humans are struggling with their own toxins, cat toxins further aggravate your stressed systems. Look to healing your digestive systems, your adrenals, your livers, and your kidneys, and your allergies will dissipate. Help clean up all of these toxins in both your bodies and the earth and we'll all be healthier.

Not usually, unless we've chosen to bond together.

A: Our ancestors were solitary. Even today, male and female cats come together for mating only. When the kittens are born, their whole world is Mother Cat. The outside world is defined in terms of how far everything is from her protective energy. She is life-giver and teacher, pure love and affection. Once she weans us, we are cut off from this source and must develop our own protection. Sometimes we will stay with a sibling if we feel especially vulnerable. Others of us choose to live in feral-cat communities for protection. In a home environment, we can learn to live with another cat *if* it serves each of us. Some of us enjoy playmates and surrogate siblings or mothers, especially if we were separated from our mothers too soon or were sick or injured and need protection. If we choose to live in a family with other cats, it's because each cat benefits in some way.

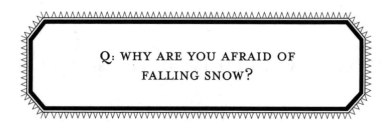

What does a desert-dweller know of snow?

A: Snow is really a foreign concept to our cellular memory. It invades our space and gets in our eyes. It's cold and wet. We love warm and dry. Sometimes kittens find it amusing for a while. If we have lived around snow for generations, it can be tolerated.

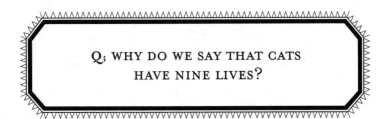

We give you a glimpse into other dimensions.

A: We are resourceful and resilient survivors. We can shift dimensions to avoid dangers when absolutely necessary. Death does not frighten us, and as a result, we are not afraid to face death at any moment and negotiate. We don't give in or give up easily.

Nine is a magic number. It indicates completion of a full circle. We all live by the numbers one through nine. Ten is a new beginning. In ancient Egypt, in Atlantis, the number nine was associated with divine inspiration because an initiate in the sacred arts had to pass through eight levels before arriving at the last, nine, in order to go to the next level of training. In these traditions there were nine steps, or lives, to go through in order to pass to another level of mastery. This belief in our having nine lives comes from the time when wise women and men understood our magic powers of healing and of avoiding death.

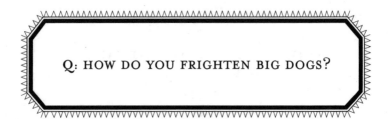

Big energy!

A: When a large animal or person approaches us, one of the ways we defend ourselves is to "blow up" our energy field so that we appear bigger than we actually are. Since dogs perceive our energy fields before they see our physical bodies, this smoke screen works to make them think that we are large and dangerous. Because dogs have strong self-preservation instincts, they usually hightail it away from us. If the dog is smart and takes a moment to really look at us, he will realize that we aren't as big and dangerous as the image we've projected. Hopefully, by the time the dog has realized this, we're long gone!

Excitement!

A: When we see a bird or a bug from the window, our hunting instinct is stimulated and we get really excited. Our whiskers twitch and our mouths quiver with the expectation of catching the bird or bug. We become completely absorbed in the prey and chitter away until we get outside to catch it or it flies away.

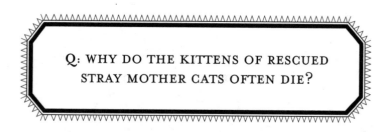

They've learned from her that the world is a terrible place.

A: Often when a stray, pregnant cat is brought into a safe place to give birth, the kittens fail to thrive, waste away, and die. Even when the mother cat is kept warm and fed, this happens. Why? Because throughout her pregnancy, the mother cat lived in fear for her life on the streets. She didn't know where or when her next meal would be. She was chased by dogs, people, cars, other cats. She passed her fears and stresses through the placenta to the unborn kittens. Before they were born, they learned that life would be filled with terror. When they arrived, they decided that it would be better to die than to live in such an environment.

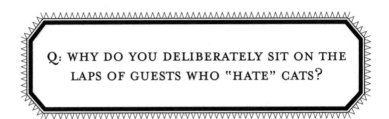

Q: WHY DO YOU DELIBERATELY SIT ON THE LAPS OF GUESTS WHO "HATE" CATS?

Because it's fun to make them squirm.

A: Again, energy has a lot to do with it. The principles of attraction and repulsion are a big part of how we work with and balance energies—our own and of the humans we choose to care for. When a person is repelled by us, we are curiously attracted to them—like a magnet! We want to push their buttons, so to speak, to see what makes them tick. When the person begins to squirm uncomfortably, we might amp up our attentions by purring and kneading their legs. We might drool with our own pleasure. What is interesting for us is to feel how these people handle our attention. Do they resist and try to push us off? Do they relax and pet us in spite of their issues? If they relax and pet us, we've actually been able to facilitate a little healing. There may be hope for them to release their tension and rigidity and walk differently through their life. If they push us off, well, they're going to stay rigid and stuck in other aspects of their lives as well.

❧

Definitely, and with deliberate purpose.

A: Our purr is how we express pleasure and reassure others and ourselves. We also use purrs of differing tones and frequencies for healing ourselves, our kittens, and our human friends. We are sensitive to the vibrations of illnesses, diseases, or imbalances. For example, when you have a cold, your respiratory system vibrates differently than when you are well. Imbalanced emotions carry different vibrations from balanced emotions. Stress affects different organs in the body, causing vibrational wobble in vulnerable places. By changing the tone, speed, and frequencies of our purr, we can help recalibrate your organs and systems to the correct vibrations.

Q: WHY DO YOU LIKE TO BE UP HIGH?

For safety and good visibility.

A: Being little and being a predator is an interesting combination. God granted us claws for climbing to help us survive. When we're higher than the bigger predators, we're safe. Even in a house, a high perch gives us a sense of security. Being up high allows us to survey the territory for big predators who might want to eat us and to spot possible unsuspecting meals.

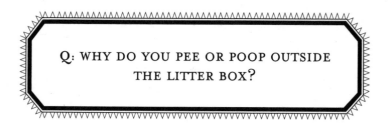

Q: WHY DO YOU PEE OR POOP OUTSIDE THE LITTER BOX?

There are many reasons.

A: First, when we are sick, we often go outside the litter box to get your attention and medical help. We stop using the litter box if it's dirty, the litter is too old, or too many cats are using the same box. We sometimes won't use the box if another cat in the house is chasing us or if a new cat, dog, or person has come to live in our house. Safety is a major concern for us, and if we do not feel safe and secure, we may go outside the box to tell you that something is wrong. We rarely do it to punish you. While we may do it to show you that we are upset about something, we are asking for your help and support. It's not a declaration of war. What would that accomplish? Almost always we would lose by being taken out of our home. This is not our goal or desire. Going out of the box is a way of getting your attention. We need your understanding and help if it becomes a chronic problem.

Q: WHY DO YOU SPRAY IN THE HOUSE?

≈

To mark territory and establish our presence.

A: It is completely natural for us to spray objects in our territory if we feel we have to defend or protect our space. All cats—from tigers and lions to house cats—do this. Safety and security are the key concepts here. Mostly we spray when other cats are in our territory and/or we feel threatened. We spray to tell the other cats that we belong here and in fact own this territory. We spray when we feel insecure or when we've lived on our own and learned to spray our outside territory.

It can be difficult for us to understand that spraying is repugnant to humans since it's a natural communication for us. Speak to us and tell us that it makes you unhappy and that you expect us to use our litter box or to go outside. Observe the other cats in the household to see if they are threatening us in any way. Check what you're feeding us. We feel safe and secure when we are nourished correctly.

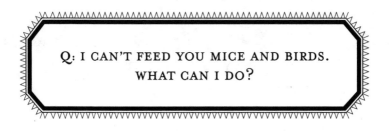

Moist food is critical.

A: Dry food is terrible for us. We get addicted to it and eventually it destroys our health. We are built to receive all our moisture from our prey. Our kidneys are not designed for processing a quantity of water. When we eat dry food, we are forced to drink water and our kidneys wear out. We cannot process the high amounts of carbohydrates in dry food. The only carbohydrates we are designed to eat are the small amounts found in the stomach of a prey animal.

Raw meat is best; cooked, homemade food is OK; canned food, if made from quality meats, is acceptable.

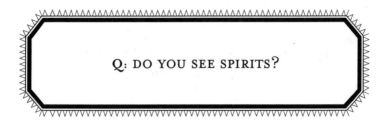

Yes, whenever they are around.

A: By spirits, we assume that you mean disembodied individuals. Because we are sensitive to energies, we perceive and even see spirits who pass through or stay in our space. Most of these beings are benign, but not all. Some are trapped between dimensions. Some carry energies that can harm living organisms. When we encounter energies of this sort, we do our best to send them away. Normally these beings are repelled by us because we are not interested in energies that are not life-affirming.

Occasionally, if we are weak or sick, a spirit can enter our bodies. This can be a problem, as we will fight the invading energy. We appear possessed because we are. If you suspect that we may be possessed, help us by feeding us good food and doing whatever you can to help us regain our physical strength. Be aware that drugs will support the invading energy by suppressing our own internal defense systems. Support our health instead with strengtheners from nature such as herbs, natural supplements, and holistic therapies. If our physical strength returns, we can usually banish the invading energies.

We also see and perceive angels. They are around us—and you—all the time offering love, comfort, and assistance. We see what you refer to as fairies or nature spirits too. Fairies are angelic or devic beings who nurture and care for every plant and flower. Each plant has its own fairy, and there are also devas of broader energy who watch over every garden, forest, meadow, etc. Many of us love to sit near plants and share in the devic energy. We communicate with these beings when we choose to. Sometimes, when fairy beings come inside to play, we watch, and sometimes we enter into the play. They are lots of fun to be with. You can observe this fairy play. It appears as if we are playing with nothing.

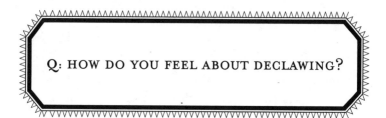

It's a horrible and traumatic mutilation.

A: When you declaw us, you cut off our first line of physical protection. On the physical level, we need our claws for climbing, to defend ourselves, and to hold our prey. On an energetic level, our claws are terribly important to us. They are a part of us, extensions of our paws just like your fingers are extensions of your hands. Claws are not just fingernails; they are part of our fingers. Think of declawing as having your fingers cut off. How would you feel?

Some of us learn to live without our fingers by rebuilding them in the physical etheric body. Some of us never recover and feel unsafe and vulnerable for the rest of our lives. This can lead us to neurotic behaviors that people find unpleasant or unacceptable. Then we are punished, banished, or killed not through any fault of our own but because you have irreparably mutilated us.

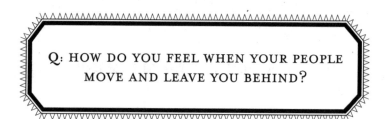

That depends on our relationship.

A: If we feel that we are a loved member of the family, being left behind is a traumatic loss for us, just as it would be for a dog or person. If we lived at the edge of the human family, we are not so traumatized. However, if we've been fed regularly by our people and suddenly they leave and there's no more food, we're in trouble. Not all of us are able to become successful hunters. There may not be a lot of prey animals available. There may be a great deal of competition for the prey that is around. Finding food and safety become our only goals. That's a lot of stress.

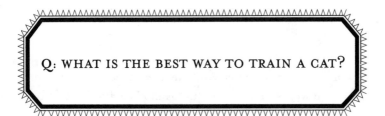

With love.

A: When we are kittens, our mother is giver of life, love, and affection. She is Source. She teaches us all about the world by speaking to us, purring to us, demonstrating what she wishes us to learn, showing us how to handle different situations. When you adopt a kitten, you must become mother. To do this, touch the kitten all over with warm, loving hands. Start at the face and stroke all the way down our little bodies. Speak to us in your softest, most loving tones. When we appear not to be listening, go to us and physically stop us to get our attention. When we are giving you our attention, tell us exactly what you want us to do. Visualize the thing you want us to do. Praise us in your soft voice and stroke us. If we have accepted you as our mother, these behaviors should be sufficient to train us.

If you adopt us when we're an adolescent or an adult, you need to establish yourself as worthy of our respect. We want you to show us that you love us, will feed and care for us, and will establish rules just as mother did. It's best to actually sit down and tell us what you want us to do in your house, and also let us know that you are unhappy when we do something you do not approve of. The best way to do this is to approach us with a clear intention in your mind and body. Your energy field will carry this intention to our energy field first. Next, your body needs to be aligned with your intention. Body language tells us a great deal. If we haven't moved by this point, physically move us if we're someplace you don't want us to be or if we're doing something you don't wish us to do. Model the behavior that you want us to do—such as play on the floor, not on the counter; play with the toy, not with the curtains; scratch the scratching post, not the couch.

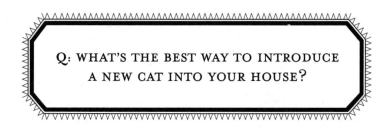

[[&]]

Carefully and with the approval of the original cat or cats.

A: As we mentioned before, we are sensitive to the energies of anybody new coming into our space. Some of us have learned how to quickly adjust our energy fields to accommodate the change. Others need preparation time and lots of practice getting used to new energies.

The first thing to do is to speak with us. Tell us that you'd like to bring a new cat (or kitten, dog, puppy, person, baby) into our shared space. Tell us why you wish to do this. Be clear and committed. If you're unclear, we will sense this and become instantly uncomfortable and on alert. Ask us to accept this new being into our home. Tell us how you will take special care to honor our process with the newcomer by watching carefully to see if we need a quiet space away from the newcomer and if we need extra attention or reassurance. Tell us when the newcomer will arrive so we can prepare. Let us sit with this news for twenty-four hours, if possible.

When you bring a new animal to the house, keep him/her in a carrier and announce to us that he or she has arrived. Take the newcomer to a private room to allow him to get used to the new space. Feed him in this room. Tell him about us and how you'd like us all to get along. Bring him something of ours to smell once he begins to express some curiosity about his new home. If he's curious and comfortable smelling the article, leave it for him in a corner of the room or somewhere it's not "in his face." Take a towel or blanket that he has used and give it to us to smell as well. Tell us again how you'd like us to accept him as a member of the family. If the newcomer is actively exploring his room, eating well, and being affectionate to you, open the door a crack to allow him to smell the rest of the house. Visualize us coming gently to the door to smell him. When we come, tell us all how happy you are that we are meeting. If this goes well, repeat it a few times until it's clear that we are all used to each other and willing to go to the next level.

When you feel we're all ready, announce to us that he will be coming out. Ask us again to be friendly and welcoming. Visualize this all going well. Open his door slowly and allow the newcomer to choose if he's ready to explore. Stay with him as he explores the house. Pet him and encourage him. When we meet, give us time to adjust to each other. If there is any hissing and the newcomer retreats to his room, that's OK. Tell us, if we did the hissing, that it is not OK to be offensive, and tell us that you expect us to do better next time. Give us at least two hours before trying again.

The key to success is patience on your part as well as constant communication throughout the entire process. Don't be anxious, nervous, or worried, or we will be too. Always be positive and clear.

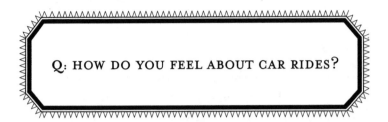

Awful, usually.

A: Car rides are unsettling because the vibrations and the changes in light and sound are usually overwhelming. We feel vulnerable and desperate for a safe space. Some of us try to get as high as possible; some of us try to get under the seat. It's always better for us to be in a carrier we can't claw our way out of. Some of us can get used to traveling in a car if we do it enough and can process all the stimuli. If the only times we go in the car are to go to the vet, you can be sure that we will associate pain and stress with car rides.

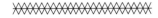

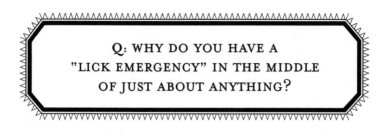

Licking ourselves is reassuring.

A: When we are excited, stressed, confused, or embarrassed, we can reassure ourselves by licking. We learned this from our mother. Don't you wish you could lick yourself to feel instantly calmer and better able to cope?

Sometimes we lick ourselves right after you pet us. We aren't trying to lick off your scent; we're trying to smooth out the energy. Often when you pet us, there's a buildup of energy where you touch us. Sometimes we bite to stop you, and other times we just lick ourselves to balance it out.

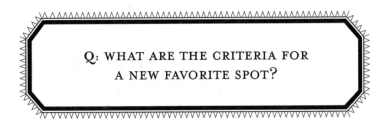

It's right, right now!

A: A favorite sleeping spot has many qualities. First, it's safe. Second, it's comfortable. Third, it allows us privacy and/or a good vantage point. A spot that is just right in the summer may be too cold in the spring, fall, and winter. Favorite winter spots—if we live in a cold climate—are usually sunny and/or warm. Most of us love the sun and really miss it if we live in an apartment that receives no direct sunlight.

If you have an apartment with no sunny spots, consider buying a full-spectrum lamp. Depending on the lamp, we could accept it as a temporary substitute. The lamp must not be too close to us but just at the right distance for us to enjoy it. Each of us will have to tell you if the lamp will be acceptable or not. Ask us where we want it located in the room and how close to our bed it should be. We'll show you, if you ask.

We change favorite spots for many reasons depending on the season, the weather, and the energies in the house. When your bed becomes our favorite spot, we are telling you that we enjoy your energy or that you need our help to recalibrate your energies.

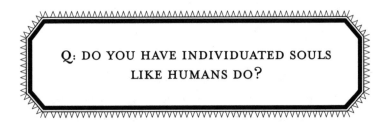

❧

Absolutely.

A: Let us state unequivocally: Cats have individuated souls just like human beings. We can't imagine why you would think otherwise! Each of us is evolving, lifetime to lifetime, just as each one of you is. We experience life, we learn, we grow. We are aware that we do this. We remember our past lives, and we know that we will be born, live, die, and be born again.

It is interesting to us that some of you question this concept. Look around you. Every living thing on planet Earth is born, lives, dies, and is reborn again. How can you think that humans do not experience this cycle? Perhaps you view it in physical terms only. Well, as the DNA of an oak tree is passed on to its acorn, so is the tree's consciousness and a piece of its soul. As the DNA of you humans is passed on to your children, so is a piece of your soul. As the DNA of a cat is passed on to our kittens, so is a piece of our soul. The soul remembers every experience it ever had. The soul remembers that it is individual and, at the same time, one with all other life forms. This is how we are all connected.

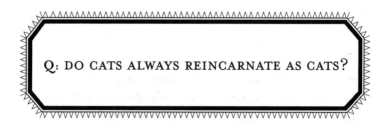

Q: DO CATS ALWAYS REINCARNATE AS CATS?

❧

Usually.

A: As beings of free will, we have the opportunity to come back and experience another life form. However, due to our relationship with energy and healing, we usually return to continue our work as cats.

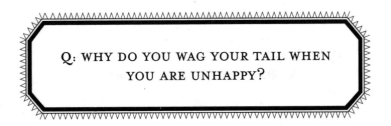

We don't wag our tails, we express with our tails!

A: Dogs wag their tails. We express a full spectrum of feelings with our tails. If you watch closely, you can learn a great deal about us from how we carry and use our tails. (Tailless Manx cats are at a disadvantage here.) We use our tails for balance, but we also use them to express our sense of self and our level of concentration, irritation, displeasure, affection, or ownership. Because our tail language is so personalized, you'll need to watch us and learn what we're communicating.

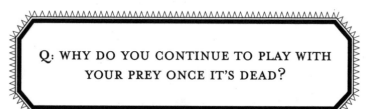

Gratitude.

A: When we have caught a mouse, for example, there's cause for celebration! We are extremely grateful to the mouse for giving us its life. The mouse has to give itself to us. It's not a competition for survival between cat and mouse. It's a dance between predator and prey. Many times the mouse leads the dance and then leaves us, suddenly, "on the dance floor without a partner," pondering the skill of the mouse. Other times we lead the dance and make a meal of our partner. This is a sacred exchange, totally understood and respected by each dancer. The mouse's death means that our life will continue. We express our gratitude to the mouse by continuing to dance. But this may look like playing with it to you.

Sort of.

A: That's an interesting interpretation. When we look at you and slowly close and then open our eyes, we are telling you that we appreciate you and enjoy your presence in this moment. If you do the same to us, we feel that you are telling us the same thing. It's more like we're *blowing* you a kiss when we close and open our eyes at you. We give you a real kiss to express our affection when we wash you with our tongues.

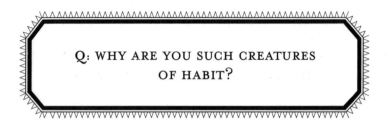

Q: WHY ARE YOU SUCH CREATURES OF HABIT?

Habits can provide security.

A: We feel safest when we have a routine, providing us with a sense of equilibrium, structure, and clarity. Energy patterns fluctuate in more predictable ways when we do the same things at the same times each day. We are not merely repeating things out of a lack of creativity. We are very aware that we are choosing to follow the same patterns.

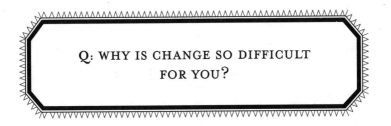

It causes a disruption in the energy flow.

A: Change can be felt as threatening. Anything new may be dangerous because it upsets the energy patterns around us. For example, let's say you buy a new couch. First, strangers come into our space and remove the old couch. We have to process and adjust to their energies. The size, shape, and smell of the old couch were familiar to us. We allowed it into our energy field and built our energetic comfort zone around it. Now, suddenly, it is removed. A big hole exists. We feel vulnerable. And then, before we can process the lack of the familiar couch, a new one is brought in. It smells different. It looks different. Its dimensions are different. Its energy is different. We have to recalibrate and adjust. Most of us do learn to adjust and create a new comfort zone. It takes each of us our own time. Some of us, due to our high level of sensitivity, take longer.

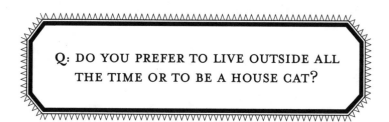

Personal preference is a factor, but we want to be with humans.

A: Thousands of years ago, we made a commitment to live with human beings. We deliberately linked our destiny with yours. In the beginning, we had complete freedom to come and go as we wished. Some of us still have this luxury and would rather die than give it up. Others of us are happy to have a safe, secure home, loving people, and regular meals. Those of us who choose to spend a great deal of time with our human beings are usually working to help heal and balance them as well as to develop ourselves spiritually.

Some beings choose to be born as feral cats, living a challenging life between wildness and human society. Theirs is a "transition" lifetime, as many beings who are born feral are new to cat form and have been wild creatures in previous incarnations. They choose to incarnate as feral so they can learn how to live in a cat body and experience feline reality, step one. These beings will never be comfortable in a house. Often these beings will incarnate two or three times as a feral cat. In each successive lifetime, they will attempt more contact with human beings until they are ready to be born to a mother who lives with humans. Sometimes, an experienced cat soul is born into a feral community. These beings will usually seek out people and allow themselves to become house cats.

Each of us chooses the best path for himself.

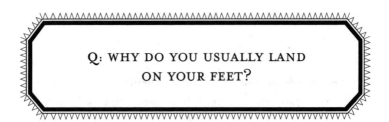

Great balance.

A: Our bodies are built to land in the most efficient way possible for survival and escape. The distance we fall determines if we can land on our feet. If we fall from too high or too low a place, we won't land on our feet.

On an energetic level, always landing on our feet tells you that we are never caught off guard. We are always prepared, resourceful, and resilient. We role model this attitude for you.

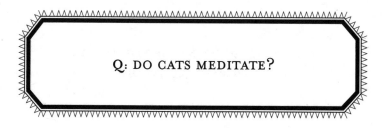

Q: DO CATS MEDITATE?

Of course we do. Shouldn't you?

A: It is important for us to become still at times to listen to our heartbeats, to connect with All That Is. Many times each day, we will sit with paws turned under or in the sphinx position, tuning in to the divine energy within and without us. When we meditate, we scan our own bodies for imbalance. We become fully present with all the processes in our bodies and energy fields. We practice full awareness.

You can learn this from us if you don't already know how. The next time your cat sits to meditate, join her. You'll learn that you are much more than you or most other human beings think you are.

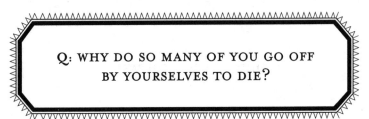
Dying is a dance, and God is our partner.

A: Dying is a process of releasing the physical body and reconnecting one's essence to God. When we approach the process alone, we enter into an intimate moment with God. We dance with God as our spirit leaves our body and is reabsorbed into the Light. It is a very private moment for many of us.

It is in our nature to do most things alone. We enjoy and appreciate our own company! It is more natural for us to go off to die than it is for us to seek out human companionship when we have to leave. When we do seek out human companionship in our final hours, it is most often because our human needs to be with us, to accept our passing, to acknowledge it, and to begin his or her own process of letting go.

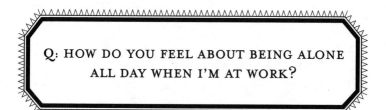

Fine.

A: Young kittens do not enjoy being alone for long periods until they are about one year old. It is not natural for them to be alone at a young age. On the other hand, mature cats are fine on their own as long as they are comfortable in their space. Because we are nocturnal, our natural schedule would be to sleep during the day and be most active in the evening, night, and early morning. However, when we adjust to your schedule, we'll be most active in the morning and when you come home in the evening until you go to bed. We will need some quality time from you when you're home, or what's the point in living together?

Q: HOW DO YOU FEEL ABOUT EUTHANASIA?

Sometimes it's helpful; often it's invasive.

A: Euthanasia can be helpful if we are trapped in seriously sick or injured bodies. Usually we would choose to die on our own, but we realize that most of you are not comfortable watching us die. Of course, you don't have to watch. You could just leave us alone. This takes a great deal of detachment, and we understand that detachment is hard for many of you.

Euthanizing healthy cats and kittens is murder. However, if you must be a part of this practice, it would help us immensely if you spoke to us before and allowed us time to prepare our souls for departure. If we have time to prepare, we can go more easily and peacefully.

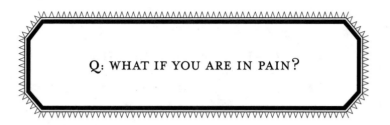

Q: WHAT IF YOU ARE IN PAIN?

≈

Pain is personal.

A: If we are struggling or in great pain, you will know.
Sometimes you are in more emotional pain than we are in
physical pain. Choose euthanasia when there is no other
way. Know that if you rush us, we will have to do more work
in spirit to clear the memory of inappropriate euthanasia
or else carry that trauma into our next life.

Dogs bring unconditional love to humanity.

A: On a spiritual level, dogs are our brothers and sisters. They have chosen to accompany human beings on the journey, just as we have, just as horses have. Dogs have committed themselves 100 percent to human beings. This is a great gift to you. We deeply honor and respect the beautiful gifts they give.

On a physical level, dogs are big predators. They are pack-oriented, which spells danger for small, solitary felines. Unless we grow up with a dog, it isn't easy for us to get along. We trigger their predator instincts just as they trigger our fears. If we know that we can find a high, safe place, we enjoy playing tricks on dogs. They can be entertaining when we're not scared to death of them.

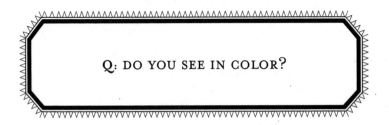

Q: DO YOU SEE IN COLOR?

We perceive color on many levels.

A: We see color as you do, but in more intense hues. Because we perceive the vibrations of color, each color feels different to us. For example, if you wear a red silk dress, we will feel a rapid vibration; if you put on a blue silk dress, we will feel calmer and more peaceful.

We perceive non-physical colors as well, such as the different colors in your energy field. Your energy field carries colors that indicate balance, imbalance, sickness, and wellness. A body enjoying vibrant health has different colors than a diseased body. Some humans know how to perceive these colors as well. They, like we cats, can support your healing and rebalancing if they choose to.

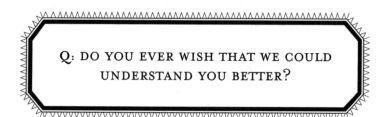

Q: DO YOU EVER WISH THAT WE COULD UNDERSTAND YOU BETTER?

Understanding is key to our reason for being with you.

A: The point of this book is to give you insights and understanding into who we are and how we are communicating with you all the time. We've offered many tools to help you understand us better: watch our body language, particularly our tails, eyes, and ears; feed us well; provide a safe, comfortable home for us; be more sensitive to our energy fields; pay attention to what agitates us and what gives us pleasure; and learn to meditate with us.

To understand another being means to be quiet and respectful, to listen, and to watch carefully. For most of you, the impediment to understanding us is a matter of time. Your perception of time keeps you always doing, running, and thinking. Few of you ever take time to just *be*. We are expert *be*-ers. You can understand us better if you take the time to *be*. Understanding us will be easy if you take the time to patiently listen. This seems to be your great challenge.

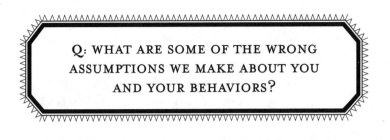

You're in trouble when you assume that we think like you do!

A: We are quite different from human beings. When you're interacting with us, it is important to look at situations from our perspective whenever possible:

1. *Remember our size. We are constantly aware that we are prey to larger predators. This fact governs our instincts and reactions.*
2. *We like to climb and to have access to high or hidden places. This is important for our sense of safety as well as our comfort. We need to climb to exercise our claws and entire bodies.*
3. *We are solitary beings. It takes a great deal of effort for us to live in groups of any sort (cats, humans and cats, humans, cats and dogs, etc.).*
4. *We are obligate carnivores. We must eat fresh meat in order to thrive.*
5. *We are energetic beings, meaning that we are aware of and working with energies invisible to the human eye.*

6. *We love with* detached love, *which humans often describe as* independence. *Detached love is the basis of our ability to offer balanced healing.*

7. *We are beings of free will—individuated souls, just like human beings.*

8. *We know that we will live, die, and be reborn again.*

If you become aware of these basic facts, you'll make fewer errors in understanding us and greater strides in working with us to find common ground.

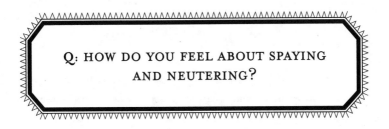

That depends on how and why it's done.

A: Sex for us is for procreation only. It isn't fun. It's a necessity. Heat cycles are challenging for a female cat. The huge hormonal fluctuations put a lot of stress on her body. Having kittens isn't fun. It's a necessity. However, we love being mothers. There's the paradox. We love caressing our babies, purring to them, watching them explore the world, teaching them how to be safe and happy. But we don't want to bring kittens into a hostile environment. Like all mothers, we want our kittens to have a chance for a worthwhile life experience. If that is impossible, we feel it is better for them not to come.

We recognize that at this time many more kittens are being born than will have homes. Life on the street is stressful at best, and most of the time it is terrifying. Spaying can be a blessing. It is a relief for us not to have to go through heat cycles.

Neutering is acceptable to most of us. If we remain intact, most of us males cannot resist the scent and energetic pull of a female in heat. We become slaves to our hormones and are often obsessed with territorial boundaries. This sidetracks us from our spiritual work.

The timing of neutering or spaying is important. When we are neutered at about six months, we will often become more loving with our human family members and more interested in non-feline relationships. Neutering or spaying us too soon, sometimes as early as six weeks old, disrupts our development, both physical and energetic.

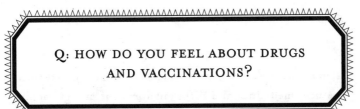

*Mostly they invade our bodies and cause
more challenge than benefit.*

A: Most of us are physically compromised in one way or another, thanks to invasive human medicine and processed foods. With more illnesses and diseases as a direct result of the chemicals we are bombarded with, we are not living as long as we did ten years ago. Vaccinations are poisoning us. Receiving them when we are kittens without fully developed immune systems damages us from the start. Giving more vaccinations to us yearly breaks down our bodies and kills us, sometimes rapidly, and if not rapidly, slowly.

Antibiotics can be helpful *if* we have a virulent infection or are unable to heal on our own. If given for other reasons, they simply get in the way of our own healing process. Steroids destroy the liver and suppress healing. Antidepressants designed for humans create distortions in our energy fields and suppress our healing as well.

Remember, we are sensitive healers. We can cooperate with natural medicines and different forms of energy medicine. These modalities enhance our own healing toolbox. Chemicals prevent us from using our healing tools.

The only true healing is reminding a being of perfect health and balance. Anything else is invasive.

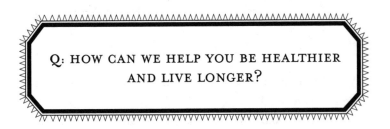

Feed us better food and take better care of yourselves.

A: Stop polluting our bodies and yours! Pollution comes at us in many forms: poor-quality food; drugs and chemical medications; chemicals in your carpets, furniture, walls, and air; discordant noises; stressful lives (yours); unhappy relationships. If you take the time to feed us correctly, we will be healthier and live longer. Look carefully at the chemical and noise pollution in our shared environment. It doesn't just affect us. It adversely affects you as well.

Stress and unhappy relationships shorten lives. Why stick around if life is unpleasant? Your body knows this. When you bring home your stresses and don't deal with unhappy relationships (past or present), we use our energy to help you. We are committed to doing this, but the more stress you have and the less you take responsibility for it, the more we use our own energy and physical resources.

We take your stress and disease into our own bodies to lighten your load.

So if you start today to take better care of yourself, we can use our energy to help ourselves deal with all the stuff we encounter living in a polluted world. Once you start to feel better and less polluted in your own life, you'll be able to address the pollution of your community, state, country, and world!

To clean and organize our fur and our energy field.

A: Licking ourselves reassures us, just as our mother did. Cleaning is a ritual for removing any foreign object from our fur that could pose a physical or energetic problem. As predators, we can't have food hanging on our bodies or our potential prey will smell us coming. We don't want our scent to attract a bigger predator either. On an energetic level, cleaning our fur combs and organizes it for maximum energetic flow. Smooth fur allows the energy to flow over our bodies much more easily and cleanly.

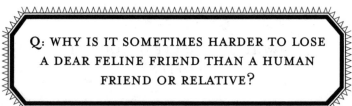

Passion and commitment.

A: We love you differently than most humans love you. We see the best in you. We are committed to helping you be healthy and whole. On some level you feel this and know what a gift this is. Our love is pure and uncomplicated, unlike most human relationships. So when we leave you, you often feel a deep pain because our love touched you so profoundly. Also, we hold in our beings a memory and reminder of your unencumbered, clear self. We know who you really are—a person of love and light beyond fears and destructive patterns. When we leave you physically, you're left behind without this loving anchor of support.

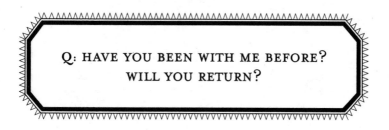

*Many of us have been with the same human being
lifetime to lifetime.*

A: Many of you know that you've spent lifetimes in the company of cats. You feel a deep connection with us and our feline ways. You have a cellular memory of working side by side in partnership with us. We have guided and supported you along your path of healing and service.

We can reincarnate with you again and again, if it's for our mutual highest and greatest good.

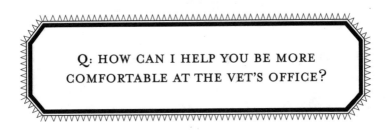

Q: HOW CAN I HELP YOU BE MORE COMFORTABLE AT THE VET'S OFFICE?

Prepare us and try to eliminate whatever stresses you can.

A: Most vet offices are challenging environments to us. We don't like being around unfamiliar animals and people. We smell and sense all that happens there. Here are some ways to help us:

1. *Take us to a clinic just for cats.*
2. *Prepare us for the visit by telling us what to expect.*
3. *Give us at least a few hours notice.*
4. *Be positive and upbeat.*
5. *Put us in a carrier we can't break out of in the car.*
6. *Put a familiar towel, blanket, or shirt of yours in our carrier.*
7. *Stay with us in the exam room.*
8. *Ask the vet to wash his hands with a non-medicated soap before touching us.*
9. *Give us no more than one medication per visit, if possible.*
10. *Bring us home as quickly as possible.*

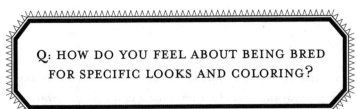

That depends on the reasons.

A: Our natural coloring camouflages us in our wide varieties of habitat. We developed longer coats in cold climates. Unusual color combinations evolved depending on our relationship to human society. For example, the slate-gray Chartreux breed developed in and among the gray rocks of Chartres Cathedral. In other words, coloring became less a matter of survival than one of beauty as we began living with humans.

Committing ourselves to humans means that we have given away some of our autonomy. We find ourselves in a predicament in certain "breeds." Most cat breeds are just variations on a main theme, but a few disrespectful breeders have created looks that actually distort and weaken us. For example, the popular Persian with a flat face is miserably uncomfortable. When we have no space for our nose and when our eyes are huge, the energy flowing to and around our face is distorted. We develop chronic secretions in the eyes and nose. Our immune systems are on constant alert. We don't want to groom because our mouth and tongue are misshapen. Persians are weak and, as a result, develop many physical and behavioral abnormalities.

If you love us, breed only for characteristics that enhance the temperament, health, and longevity of the cats in your care.

Q: HOW DO YOU FEEL ABOUT CAT SHOWS?

Cat shows are for people, not for cats.

A: You love to show us off! Some of us are OK with this. Others hate being put on display and manhandled by judges. It's a matter of personal preference. Ask your cat if he or she enjoys going to shows. Listen and watch how your cat handles the show scene. If we get used to shows when we're kittens, have a safe and secure private space, and enjoy people, shows can be OK. If you respect and honor the choices of your own cats, shows will be more pleasant for everyone.

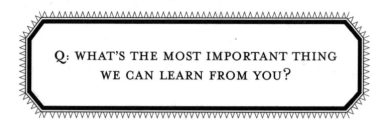

Q: WHAT'S THE MOST IMPORTANT THING
WE CAN LEARN FROM YOU?

Detached love.

A: Our greatest gift to humanity is the quality of detached love we model for you every day. Detached love is love without expectation or obligation, given without any need for receiving. Detached love is balanced and whole all by itself. Cats are great healers because we hold this quality of detached love. It allows us autonomy and wholeness. Once humanity embraces the quality of detached love, the world will be full and complete. This is where you are evolving to.

We hold this quality because we work constantly with divine energy, drawing it down and holding it in our bodies. We're experts at detached love for a reason beyond the balance and wholeness it gives us. We do this for you. Much of humanity has become disconnected from Source, from God. Your belief system is that you are separate from other human beings, from animals, from plants, from minerals, and from our Mother Earth. This is wrong. This belief system has brought you only grief and loneliness. For centuries you tried to medicate these feelings by accumulating power over others. You tried to fill your emptiness with physical objects. Many of you are now finally realizing that you have been worshipping a false god. We cats have been patiently holding divine energy for you so that when you're ready to go home to God, we'll escort you all the way.

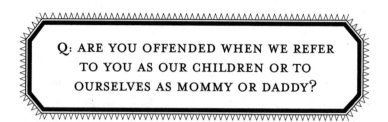

Q: ARE YOU OFFENDED WHEN WE REFER TO YOU AS OUR CHILDREN OR TO OURSELVES AS MOMMY OR DADDY?

That's a matter of personal preference.

A: We accept that sometimes you love us as your children. This is a great honor, for we love our kittens beyond description. We let you know when you baby us too much. We let you know if you're not treating us as we wish to be treated. Cats are rarely codependent!

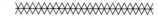

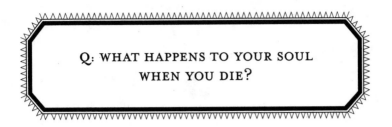

Q: WHAT HAPPENS TO YOUR SOUL WHEN YOU DIE?

We reenter the place of reunification.

A: We go to heaven just like you do. When our soul leaves the body behind, it is immediately attracted to the Light. Like iron to a magnet, the soul is drawn into the Light and reconnects to All That Is, to God. We merge our essence back into love. We feel our individuality while at the same time we feel no separation, only oneness. Our experience is that there is one place of reunification for all beings. There isn't one heaven for people, one for cats, one for dogs, one for wild animals, etc. When we pass out of form, we all reconnect, surrounded and interpenetrated by Divine Love.

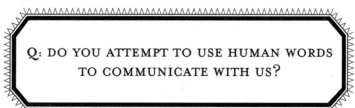

It sounds that way, doesn't it?

A: We cats know that humans respond to certain sounds.
For example, there are some universal sounds among
mammals that trigger the mothering instinct. We know
how to make sounds that are familiar to either your conscious or unconscious brain. Many of us attempt to mimic
human words to get your attention, to make you think, and
to help you realize just how intelligent we are. Yes, we know
exactly what we're saying.

We always have a reason.

A: As a predator, we have to be careful about vocalizing. If we're too loud, we'll scare away our potential meal. We have different types of howls. Females howl when they are in season. Toms howl to scare off opponents. Among ourselves, we howl in warning, to threaten, to establish territory. When we live with you, we howl to ask for food or something else we desire as well as to express displeasure or discomfort. Sometimes we just like the sound of our voice. Many of us are "talkers." Others of us hardly speak at all. It's an individual choice. We use sound frequency deliberately, and we often use it to rebalance the energies in a place.

Siamese cats developed a distinct and highly refined voice that serves us in heavy jungles amidst the cacaphony of many other animal vocalizations. Although many human beings find the Siamese voice unpleasant, we have learned to use it effectively to get what we desire.

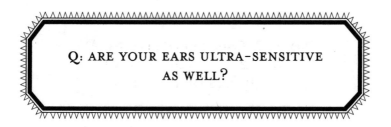

❧

Compared to yours, yes.

A: Our sense of hearing is about seven times more accurate than yours. We hear with our physical ears as well as with our third or clairaudient ear. Because we can turn our ears, we can pinpoint the source of a sound more accurately than you can. As a result, we are sensitive to loud noises, including loud, discordant music. Sound has the capability to heal. It also can disturb, distort, damage, and even kill. Some sounds you've gotten used to are damaging you. We do our best to avoid these sounds, if at all possible, because when we are exposed to loud and destructive sounds on a regular basis, our immune systems will be compromised, just as yours will be. A rule of thumb: If we avoid or hide from certain often-heard sounds in our shared environment, you can be sure that we're all in for sickness at some point.

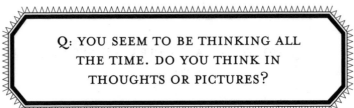

Both and more.

A: Although our brains seem smaller and simpler than yours, we think with our entire being. You have really missed so much by measuring intelligence by brain size. Look at it this way. First, humans have determined that you use 10 percent of your brains, yet you argue that because the human brain is one of the largest, you must have great intelligence. Then you go and judge our smaller brains as inferior. Has anybody ever stopped to determine how much of our brain we use? We use a lot more of our brain than most humans do!

Next, you measure intelligence based on your ability to speak, write, build things, and control the world. You're confusing a developed intellect with intelligence. We cats define intelligence as the ability to utilize all one's gifts and faculties for balanced results. This includes using one's heart and intuition as well as all one's senses, feelings, and physical and non-physical experiences. In our experience, the brain is only one component of intelligence. In human experience, most cultures' glorification of the brain has created imbalance and a great lack of intelligence. Decisions based on reason alone are lopsided and incomplete. You have only to look at the state of the planet to appreciate how this overemphasis on intellect has wreaked havoc! When humans value the other senses on the same level as the brain, true intelligence will return.

We have thoughts like you do. We perceive visual, auditory, and kinesthetic stimuli both in the physical and non-physical realms.

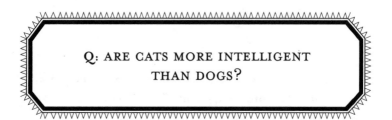

Humans are always comparing and missing the point.

A: We are not more intelligent than dogs. Our intelligence is based on our physical and non-physical experience, just as a dog's is. We are different on many levels. As a result, we make decisions based on different perceptions, different realities than dogs do.

Each cat and dog has his or her own level of understanding and awareness, just as each human has. Some cats are younger, less-experienced souls focused on survival and incapable of seeing the big picture. Others are beings who have been cats in high service for centuries. They understand more than most sleepwalking humans. The range in dogs is virtually the same. We are all growing and evolving at our own individual rate.

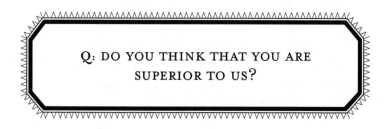

Some of us do.

A: Truly, it is not about who's better, smarter, stronger, or more successful. It's about awareness. At a soul level, we know who we are, why we are in form, and why we live with you. Many of us live with people who are completely asleep; others of us live with those of you who are awake occasionally; and others live with humans who are just waking up. Some of us are fortunate to live with awakened humans. We accept all of these experiences as valuable.

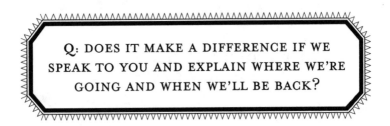

Q: DOES IT MAKE A DIFFERENCE IF WE SPEAK TO YOU AND EXPLAIN WHERE WE'RE GOING AND WHEN WE'LL BE BACK?

Yes!

A: We deeply appreciate it when you take the time to tell us what's going on. We appreciate it when you give us a little notice of an upcoming change or event as well so we have time to decide how we'll handle it. It's helpful when you tell us when you'll be home each day. If you have a clear concept that, say, 6:00 P.M. is in the evening and you tell us you'll be home at six, we'll know what you mean. It helps us a great deal when you explain why we are going to the vet's. If you do this, we will have time to shore up our energy field to better cope with the other animals at the clinic. Knowing when we're going, why we're going, and when we're scheduled to return home will positively affect our recovery from surgery and help us focus on our healing.

Tell us when you'll be traveling. We hate it when the suitcases appear and there's no explanation. And please keep in touch with us when you're away. Just make a

mental picture of us in our favorite spot. Send us a mental picture of you having a good time and tell us when you'll be coming home. We'll be much happier, and you'll be happier too!

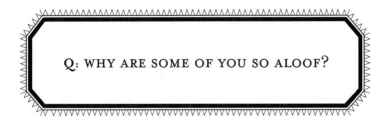

Q: WHY ARE SOME OF YOU SO ALOOF?

It's a matter of personal style.

A: Our detachment often manifests itself in what you refer to as aloof behavior. Our sensitivity to energies often requires that we maintain ourselves at a comfortable distance from others. Our choice to be friendly is related to how well we process the energies in your field and environment. Our health is also a consideration. If we are not strong and healthy, we will not be able to handle a great deal of human energy. Generally, you'll find that cats who are eating dry or poor-quality foods are often aloof. They don't have the internal resources required to process challenging energies. When we're hunting our food or eating raw meat, our bodies and energy fields are strong and we can enjoy human energy much more.

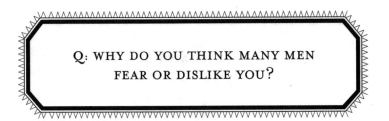

Q: WHY DO YOU THINK MANY MEN FEAR OR DISLIKE YOU?

Men who don't know us often fear us.

A: Let's face it, we're mysterious. We love the fact that humans find us to be mysterious as long as it doesn't turn to fear. We have a natural affinity for intuitive, sensitive people, and at this time in human evolution, there are more women than men who fit this description. Men who put up facades, who have denied their feelings, and who are attached to being in control can feel that we see right through them. We do. Men can feel that we are scrutinizing them, and it can make them squirm. They express their discomfort by projecting stuff onto us instead of looking at their own lives. Beware the man who dislikes cats. He often harbors a dislike or distrust of his own feminine side and possibly dislikes or distrusts women too. A man who loves cats is often in touch with his sensual, feminine side. If this is balanced, he can be a wonderful companion or mate.

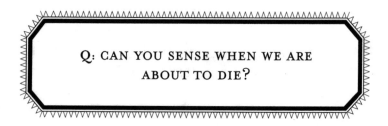

Absolutely.

A: When a body is dying, its vibrations or energetic signature changes. For many, the energy throbs and fluctuates, as if it is breathing. The spirit becomes brighter and fuller. As death approaches, the body becomes quiet, energetically, and the spirit becomes more active. When the time is right, the spirit breaks free of the body and passes to the Light.

Cats have the ability to act as midwife at a person's (or animal's) death. If we choose to be with a dying person, we will sit at an appropriate distance—sometimes close to or touching the person, sometimes apart. We will meditate and gradually enter near the energy field of the dying one. Here we will send waves of love and support, flowing with the fluctuations of his energy field. When the spirit is about to break free, we stand ready to accompany it to the Light. If the spirit needs or wishes to have our help, a part of our spirit will lead the human spirit part way to the Light. When we see that the spirit has completely left his body behind, our traveling spirit returns to our own bodies and we continue as before.

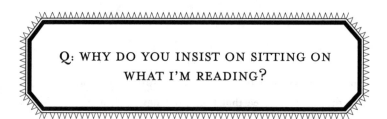

Q: WHY DO YOU INSIST ON SITTING ON WHAT I'M READING?

We love that quiet, focused energy.

A: When you are reading something, your usually active body is at rest. You are focusing your energy on the book, newspaper, or magazine. You create an energetic bubble around yourself and the book that is really attractive to us. For many of you, the *only* time you sit quietly is when you're reading, and we jump on the opportunity to share the quiet moment with you. If you're reading a newspaper, its texture and sound are especially appealing to us.

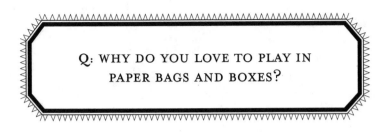

Don't you enjoy hide-and-seek?

A: Paper bags and boxes are wonderful places to play! The bags rustle and move, which is exciting. The bags and boxes provide a great place to hide out and wait for prey to come our way. We know that it's play prey, like your foot, a toy, or your hand scratching the outside of the box, and it's lots of fun. The anticipation of what might happen is just as exciting as something really happening.

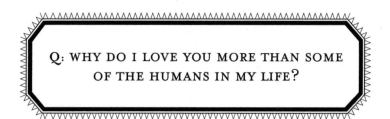

It's easier.

A: We love with clarity and accept you as you are. Our love is less complicated than human-to-human love. We don't love because we need to be loved in return. We *know* that we are loved. We're with you to help you feel connected to divine love, which is the purest of all. And at this moment in time, animals are closer to that understanding than most humans. However, those of you who have opened your hearts to us can feel our connection to divine love, and you long to feel it yourselves. We can be your bridge.

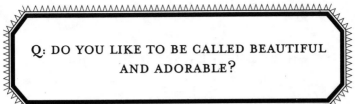

Q: DO YOU LIKE TO BE CALLED BEAUTIFUL AND ADORABLE?

Of course! We are.

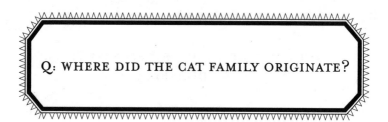

Q: WHERE DID THE CAT FAMILY ORIGINATE?

Our soul group originated from God, just as yours did.

Q: HOW WILL THE ENERGY OF THE NEW MILLENNIUM AFFECT INTERACTIONS BETWEEN CATS AND HUMANS?

≈

Very positively!

A: This is an exciting time to be alive for all of us. The earth herself is moving through a large shift—a graduation of sorts. Awareness on the planet is quickening. More and more human beings are waking up from deep slumber. Communication is bringing people together. Important questions are being asked. You are beginning to realize that you are not alone but are loved and supported by your fellow earthlings.

We know that soon, humans in great numbers will break out of the shell of separation and embrace their connection with all living beings. You are beginning to see how every thing, every system, every community, every creature is linked. You are creating links among yourselves in order to share, learn, and grow.

You are feeling how powerful it is to work together for the good of all. One by one, you are learning that you can make a difference by standing together and taking action to create community.

You are supported by the heavens as never before to achieve your true selves. As this process gains momentum, you'll learn more and more about who all animals really are. As this understanding grows, you will respect us, honor our gifts, and care for us as we need to be cared for. We will all live in greater harmony, health, and balance. The world is literally transforming before our eyes. It's good. It's time. It's perfect.

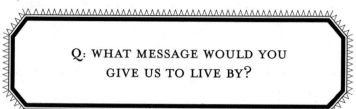

Q: WHAT MESSAGE WOULD YOU GIVE US TO LIVE BY?

Accept love. Engage fully in life. Be happy.

In Conclusion

≫

I would like to thank the many friends, clients, and "cat people" who sent me questions for the cats to answer.

Beyond Words and I would like to invite our readers to write in with questions for the next book in our series, *Conversations with Horse*. Let us hear from you!

Please send questions to

Conversations
c/o Beyond Words Publishing, Inc.
20827 N.W. Cornell Road, Suite 500
Hillsboro, OR 97124-9808

For more information on how to help our cats be their healthiest in body, mind, and spirit, see *The Holistic Animal Handbook*.

Kate Solisti-Mattelon, internationally known speaker and teacher, has been communicating telepathically with animals as a professional animal communicator since 1992. During this time she has worked with holistic veterinarians and individuals, assisting in solving behavioral problems, understanding health concerns, healing past traumas, and facilitating understanding between humans and non-humans. Kate has worked with wild animals, learning about their interrelationships and their roles in balancing the Earth. She has provided advisory support at Fossil Rim Wildlife Center in Texas. She was the guest speaker at the British Association of Homoeopathic Veterinary Surgeons meeting and the Rocky Mountain Holistic Veterinary Association meeting. Kate has been a presenter at Whole Life Expos throughout the western United States. She also presented a workshop at the International Society for the Study of Subtle Energies and Energy Medicine Conference (Boulder, Colorado). Kate has taught interspecies communication workshops in New Mexico, Colorado, Washington, New York, New Jersey, France, England, and Belgium.

Kate and her husband, Patrice Mattelon, co-authored *The Holistic Animal Handbook: A Guidebook to Nutrition, Health, and Communication* (Beyond Words, 2000). It was simultaneously published in German as *Spirituelle Partnerschaft mit Haustieren*. She has written articles for *Tiger Tribe*, *Wolf Clan*, *Best Friends*, and *Species Link* magazines. In 1997, Ms. Solisti-Mattelon co-edited an anthology, published in German, *Ich spürte die Seele der Tiere*. It was published in English as *Kinship with the Animals* by Beyond Words in 1998. Contributors include Jane Goodall, Michael W. Fox, Alan Drengson, Michael Tobias, Linda Tellington-Jones, and Michael Roads. She has been featured in *Animals: Our Return to Wholeness* by Penelope Smith and *Communicating with Animals: The Spiritual Connection between People and Animals* by Arthur Myers. Kate is also the author of *Conversations with Dog* (Beyond Words, 2000).

Together, Kate and Patrice bring the unique aspect of working as a married couple in the field of holistic health for animals and people. Since the summer of 1996, they have been helping human beings understand and nurture the physical, emotional, mental, and spiritual aspects of

the animals in their lives for the greatest mutual benefit
and growth. They have been featured in *Pets: Part of the Family*
magazine (July/August 1999) and in newspapers and on
television and radio. In the Denver and Santa Fe areas they
teach ongoing classes on holistic animal care and holistic
human growth. Kate and Patrice live in Boulder, Colorado,
with son Alex, daughter Miranda, and cat Azul. You can
visit their Web site at *http://home.earthlink.net/~solmat*.

Conversations with Dog

An Uncommon Dogalog of Canine Wisdom

Author: Kate Solisti-Mattelon

$13.95, softcover

   Conversations with Dog is a groundbreaking book in the field of human-animal communication. In a question-and-answer format, this is the first book of its kind to pose questions to dogs and receive answers in return. These answers are not based on what human beings suppose dogs think. Instead, Solisti-Mattelon, a practicing animal communicator, goes straight to the dogs themselves. Most of us who own dogs know they are trying to tell us something, and *Conversations with Dog* breaks through the species barrier to ask dogs what they want to tell all of us.

The Holistic Animal Handbook

A Guidebook to Nutrition, Health, and Communication

Authors: Kate Solisti-Mattelon and Patrice Mattelon

$14.95, softcover

The Holistic Animal Handbook is the first book to bring together practical information about diet, nutrition, and training with animal communication and emotional balancing techniques. The book is the result of the authors' many years of experience working with companion animals and their people. It includes chapters that explain how to prepare healthy, holistic recipes and Bach Flower Remedies for restoring an animal's emotional balance. There is also a chapter that describes natural techniques for dealing with common behavioral and training problems. The goal of *The Holistic Animal Handbook* is to provide animal guardians with a starting point from which they can foster and practice deeper interspecies communication. Focusing primarily on dogs, cats, and horses but relevant to virtually all animals, the book presents a dual premise: healthy companion animals are better equipped to help the humans they love, just as educated humans are better able to comprehend what their animals are about.

Kinship with the Animals

Editors: Michael Tobias and Kate Solisti-Mattelon

$15.95, softcover

Contributors to *Kinship with the Animals* represent a myriad of countries and traditions. From Jane Goodall illustrating the emergence of her lifelong devotion to animals to Linda Tellington-Jones describing her experiences communicating with animals through touch, the thirty-three stories in *Kinship with the Animals* deconstruct traditional notions of animals by offering a new and insightful vision of animals as conscious beings capable of deep feelings and sophisticated thoughts. The editors have deliberately sought stories that present diverse views of animal awareness and communication.

Other Books from
Beyond Words Publishing, Inc.

When Animals Speak

Lessons, Healings, and Teachings for Humanity
Author: Penelope Smith; Foreword: Michael Roads
$14.95, softcover

This book offers deep, life-changing revelations, communicated directly from the animals. Discover who animals and other forms of life really are; how they understand themselves and others; how they feel about humans and life on Earth; how they choose their paths in life and death; the depth of their spiritual understanding and purposes; and how they can teach, heal, and guide us back to wholeness as physical, mental, emotional, and spiritual beings. Regain the language natively understood by all species. Laugh as you experience other species' refreshing and sometimes startling points of view on living in this world, among humans, and with you.

Animal Talk

Interspecies Telepathic Communication
Author: Penelope Smith
$14.95, softcover

If your animal could speak, what would it say? In *Animal Talk*, Penelope Smith presents effective telepathic communication techniques that can dramatically transform people's relationships with animals on all levels. Her insightful book explains how to solve behavioral problems, how to figure out where your animal hurts, how to discover animals' likes and dislikes, and why they do from gangly fledgling to Grand Goose and his triumph over the turmoils of his soul and the buffeting of a mighty Atlantic storm. In *The Great Wing*, our potential as individuals is affirmed, as is the power of group prayer, or the "Flock Mind." As we make the journey with this goose and his flock, we rediscover that we tie our own potential into the power of the common good by way of attributes such as honesty, hope, courage, trust, perseverance, spirituality, and service. The young goose's trials and tribulations, as well as his triumph, are our own.

Flower Essences for Animals
Remedies for Helping the Pets You Love
Author: Lila Devi; Foreword: Rena Ferreira, D.V.M.
$14.95, softcover

Flower essences—herbal tinctures for strength and balance—are widely used to treat human ailments and are also a simple and effective means of pet care in both daily life and emergencies. Completely safe and gentle, yet powerful, these essences activate the animals' own ability to heal itself. A brand-new concept of "theme essences" to determine your pet's innate character strengths and your own as a pet owner is included.

Dolphin Talk
An Animal Communicator Shares Her Connection
Author: Penelope Smith
$9.95, audiotape

Tales of adventure and communications from the dolphins transport us into the excitement and mystery of our eternal connection with these charismatic marine mammals. Smith shares the dolphin healing journey that enabled her to over-the things they do. Without resorting to magic tricks or wishful thinking, *Animal Talk* teaches you how to open the door to your animal friends' hearts and minds. An entire chapter of this illuminating book is devoted to teaching people how to develop mind-to-mind communication with animals. *Animal Talk* also explores the topics of freedom, control, and obedience; understanding behavior from an animal's point of view; how to handle upsets between animals; tips on nutrition for healthier pets; and the special relationship between animals and children. There is even a section on how to communicate with fleas and other insects!

Listening to Wild Dolphins
Learning Their Secrets for Living with Joy
Author: Bobbie Sandoz
$14.95, softcover

Listening to Wild Dolphins, written by a well-established therapist, chronicles her remarkable and healing experiences while swimming with a pod of wild dolphins off the shores of her Hawaiian home over the past ten years. She has observed that the dolphins have qualities which humans can model to become more balanced and joyful in everyday life.

The Great Wing
A Parable
Author: Louis A. Tartaglia, M.D.
Foreword: Father Angelo Scolozzi
$14.95, hardcover

The Great Wing transforms the timeless miracle of the migration of a flock of geese into a parable for the modern age. It recounts a young goose's own reluctant but steady transformation

BEYOND WORDS PUBLISHING, INC.

Our Corporate Mission:

Inspire to Integrity

Our Declared Values:

We give to all of life as life has given us.
We honor all relationships.
Trust and stewardship are integral to fulfilling dreams.
Collaboration is essential to create miracles.
Creativity and aesthetics nourish the soul.
Unlimited thinking is fundamental.
Living your passion is vital.
Joy and humor open our hearts to growth.
It is important to remind ourselves of love.

come a lifelong fear of deep water and swim with wild dolphins in the open ocean. Feel the special affection for the human species that the dolphins impart; hear about the merging of dolphin and human consciousness; experience the haunting tones of dolphin "resounding skull bone chanting," creating an opening in the listener's skull to better receive the dolphin's energetic transformations. According to Smith, the dolphins facilitate the weaving of energy matrices of consciousness over our planet, allowing receptive and ready humans to receive the dolphins' pure love throughout their cellular structure and to experience telepathic communication. Feel the dolphins' healing power conveyed through this audiotape, directly by them!

To order or to request a catalog, contact
Beyond Words Publishing, Inc.
20827 N.W. Cornell Road, Suite 500
Hillsboro, OR 97124-9808
503-531-8700 or 1-800-284-9673

You can also visit our Web site at *www.beyondword.com* or e-mail us at *info@beyondword.com*.